I0049768

The Amazing Biological Revolution And

The Amazing New Health Care

Can We Make It Real?

What Has COVID-19 Taught Us?

What Will The Biological Revolution Mean, And How Can We Turn Fantastic New Biological Insights Into Effective Pre-Emptive Highly Effective Health Care - That We Can Afford?

PROFESSOR BERTIL LINDMARK
MD, PhD
University of Gothenburg
SWEDEN

THE AMAZING NEW BIOLOGY

Table of Contents

Dedication

Health is the primary condition for living a fulfilling life, to become an inspiration and to provide for others. Our quest for lifelong health has been ongoing for centuries if not millennia. We live in a world of multiple ongoing revolutions in basic sciences and in applied sciences and I would like to dedicate this book to all those human beings who spend their energy and their time on this earth to find the truth through a scientific method, aiming at making life better.

Acknowledgement

This work was started at the time I was appointed a visiting professor at the Institute of Medicine and the Gothenburg University in Sweden. Since I have spent most of my career in the pharmaceutical industry and the key subject of the professorship was Innovation and Entrepreneurship, I had not done much academic writing outside of my PhD research and my postdoc science. I was thinking that I should spend some time to review where we are in our biological understanding and in the various ways we can use this knowledge to help patients. I was very pleased Dr Lisa Ooi, whom I had the pleasure to work with during my time in Singapore, agreed to help me with the cell therapy portion of the book. I would also like to acknowledge the help of team SEO PROHUB UK for editing the book and helping me complete this effort.

About the Author

Professor Bertil Lindmark is a trained physician, who worked for 10 years and achieved board certificate in Internal Medicine and in Gastroenterology. After finishing his PhD thesis at the Lund University in Sweden focusing on the molecular epidemiology in the area of protease inhibitors, he started his career in the pharmaceutical industry. He held globally leading roles in Astra and AstraZeneca, between 1991 and 2010, mainly in the area of Respiratory and Inflammation, and was instrumental in the development of the billion-dollar selling drug Symbicort for asthma and COPD. After a secondment to AstraZeneca Japan where he led the Clinical Development, he moved to become the global head of R&D at Almirall in Barcelona, Spain. There he developed the second long-acting antimuscarinic to the market in the US and in EU, and several other compounds, which were acquired by AstraZeneca in 2014. He then became the Chief Medical Officer of ASLAN Pharmaceuticals in Singapore and led the development of several oncology and dermatology compounds. In 2019, he started at eTheRNA Immunotherapies in Belgium as the CMO leading the clinical development of cancer immunotherapies using similar technology as Moderna and BioNTech.

Since 2020, Bertil Lindmark has been serving as the CMO Galecto Biotech in Copenhagen, Denmark. His focus here is to use antagonists of galectin-3 which is a central pro-inflammatory and pro-fibrotic molecule, aiming to develop medicines against fibrosis and cancer.

Bertil is happily married to Emelie Lindmark, and they have a combined family of 6 adult children and 3, and soon 5 grandchildren. Outside of the professional life

Bertil enjoys reading, cooking, playing guitar and sleight of hand magic.

1 Can You Feel the Bio Revolution – It Is All Around You!

There is a biological revolution going on around us. Every day we hear broadcasts of new findings central to disease processes. New technology and new insights keep building. But can we as individuals reap the benefits of this revolution, and can we make it accessible to the many people around the world? Can we keep the new ways to interact with our biology safe?

At the same time, we see the limitations; we are still suffering with the COVID-19 epidemic. There is not enough quality knowledge reaching us quickly enough, and disinformation is leading people astray. We are in need of rapid translation of insights into therapeutics. But surely enough, we are on a fantastic trajectory for the future.

Think back to 1920 – most of us were not born yet, but what did people know about biology and about medical therapeutics and surgical interventions? Not so much.

For several hundred years, the knowledge has been building and we have seen an accelerating pace of new insights and possible interventions, both regarding diagnostics and therapeutics. The fundamental understanding of the genes and new ways to reprogram the genetic material in live tissue using CRISPR/Cas9 and similar techniques allow imagination to leap forward at an even faster speed today.

But health care is basically still organised like it was in the 1920s. Most of our high-tech resources come into play when diagnosing a very threatening medical event has occurred, like a stroke, a heart attack or cancer. Often, much is already lost in terms of organ function and the prognosis turns out poor. What if we could turn health care on its head and instead of the passive catastrophe and angst-based current paradigm turn health care to presage and preempt deterioration of organ function and allow long, disease-free lives.

The conundrum we face is that we spend too much resource on disease states that show a lesser likelihood of improving radically and much lesser on prevention. Can we turn this around and can we get to a position where we truly understand the basis of disease in a certain person, and thus be able to provide disease-curing personalised therapy? Walt Disney said: "If you can dream it you can do it." And we believe we may get there even though it may take 50 years or more. But looking back to 1920, over our last 100 years, and now observing the multiple parallel revolutions in diagnostics and therapeutics as well as our scientific methods in general, including the IT and AI and deep learning revolution, it may take less than 100 years to get to our new dream.

As we advance 20 years, we may assume most people in the more affluent countries have had their genome sequenced, and that many will have access to sources of stem cells like umbilical cord cells from birth. Obviously, we may even detect severe conditions intra-uterine, and we may land at the difficult decision of genomic correction or abortion. Religious belief and political affiliation will add to the discussion apart from personal life choices of the parents. For persons with defined disease, the future may offer novel experimental interventions and there will be difficult choices to make whether in trying something new or sticking to what is known. In severe diseases like metastatic cancer, severe autoimmune diseases, in immune deficiency states or in severe monogenetic disease, the choice may be more obvious. However, in other instances the benefit risk and the ethics involved may make therapeutic choices harder.

But let us stay positive and let us believe we can turn around health care to become preventive and preemptive using the new insights into biology. Let us hope to avoid suffering from catastrophic medical events and severe diseases altogether. Give us 100 years more and we will move to a kind of health care which nips disease in the bud, cures it preemptively, and allows us to live without worrying about the spectre of disease.

If we cannot achieve preemptive health care, then we should go for the vision of 1 day diagnosis and same day cure!

2 The Biological Revolution

In 1955, the year when Einstein died, we knew about major organs and organ function, about cells and a bit about biochemistry. Many important medicines had been invented, and medical chemistry that once evolved as an offshoot of chemistry, invented for the textile industry cloth dying, was progressing. However, not so much was known about cellular and subcellular functions, nor about proteins, DNA, or RNA. Immunology was developing but no way near to what we now can conceive about it.

As we understand, the first humanoids that have been discovered are from 3.2 to 3.9 million years back, representing 'Lucy' (Australopithecus) and the recent Ethiopian pre-human craniae.[1]. And then think about the time period over which our parallel ongoing revolutions in mathematics, microelectronics, computer design and programming, physics, material research, biology and medicine have accelerated, as well as the expansion of the funding mechanisms and funds. Is this the beginning or the end of the beginning? One would rather believe that the sky is the limit in terms of knowledge development across all these areas. As a race, we have just scraped the surface of the vast sea of knowledge that one would believe lies before us.

So, from simple medical interventions ranging from wound treatment, simple surgery and use of plant-based therapeutics, to vaccination, small molecule medicines, Keyhole surgery, microsurgery, peptide drugs and protein drugs, biologicals, personal therapeutics including

---

[1]Barras, C. (2019). Rare 3.8-million-year-old skull recasts origins of iconic 'Lucy' fossil. Retrieved from https://www.nature.com/articles/d41586-019-02573-w

genetic therapy, cell-based interventions against cancer, and radiologically-based therapies, detailed diagnostics including imaging (ultrasound and Doppler ultrasound, computerised tomography and magnetic resonance and positive emission tomography), heart examinations (ECG and heart catheterization), blood- based examinations (including study of bone marrow, liver, kidney pancreas and endocrine organs and tumour markers), invasive diagnostics (cell-based and tissue-based with the help of pathologists and study of the muscular and nervous system (electromyogram, electroencephalogram, special examinations of the eye and ear and balance), and new medical devices like prostheses, inhalers, pacemakers, electrostimulation devices (pain control, nerve stimulation), implantable devices (for hearing, sight, against depression, against epilepsy) and transplantation technology (including organ selection and preservation to surgical techniques and immune suppression), we have come a long way. This is an ongoing diagnostic and therapeutic revolution, which builds on new understanding of biological systems on organ and cellular and subcellular levels along with taming of photons and electrons and neutrons, and adding the acceleration technologies of artificial intelligence, machine learning, simulations and modelling as well as use of big data.

So why write this book then, since all is changing so fast? Well, it may be a good idea at the time to pause for a while and reflect on where we stand. We need to allow ourselves to speculate how these evolving technologies may influence the care of a singular patient. This in turn may help to understand how health care systems need to evolve to both effectively implement and achieve real benefits from the innovations at reasonable risk levels, and also how health care systems should progress. The latter point implies utilizing the best organisational models, IT support, finding the right balance between generalists and specialists, between access to local care and hardcore specialist care, and balance between prophylaxis, primary prevention and secondary prevention. We are at the point where we are expecting health while we are growing older and whilst our ability to pay for diagnosis and therapy or cure is insufficient and decreasing. We need to occupy ourselves with this conundrum because it will determine how well we can live in the future. This book tries to tell the story of the ongoing revolutions, and asks the question: how can we best translate these new biological insights into beneficial outcomes for mankind? Your life depends on it.

3 The Best of Times or the Worst of Times?

3.1 Ongoing Epidemiological Explosions – the Perfect Storm?

3.1.1 Population Explosion

Until the 1950s, the global population of human beings was under 3 billion. We are now at 7.7 billion, meaning we have doubled in 70 years. This could be seen as something bad or something good or something in between. The image of the population explosion in the 1960s was that of famine, dangers and threats. Today the negative focus is around environmental stress and climate change. In spite of the negative imagery, extreme poverty (which is one of the most clearly related causes of disease and early death) is reducing, access to healthcare is increasing, maternal care is improving, literacy is improving, weaponised conflict kills and deaths are reducing and vaccination programs are on the whole working. These prove that maybe we are not at the abyss yet. Other risks may increase, such as risks for pandemics (recall swine flu epidemic) and rapid spreading of deadly viruses (although epidemics such as the Ebola crises have been confined locally and only sporadic cases existed outside of primary affected areas).

In the past geopolitical conflicts were settled using the mechanisms of war and physical blockade of trade or imports. Nowadays conflicts may be executed via financial means (compare Reagan's star war and the fall of the Soviet Union), or via data or via disinformation destabilization (compare US election 2016 and Brexit). Still, the consequences may be in the form of worsening economics and this will have health consequences.

3.1.2 Fat Explosion

The world is getting fatter. This is best illustrated by the overweight maps of the USA. The causes may be multiple, such as general access to food and calorie density, size of portions, carbohydrate content, use of high fructose corn syrup as a basis for food production, sedentary lifestyle, less physically active jobs and reduced smoking (smoking increases calorie burn rate by about 10%). The fat explosion will lead to increase in all types of disease, including cardiovascular, musculoskeletal, cancer and infections, renal disease, hepatic disease, endocrine disease (read diabetes), cerebral cognitive disease. So, from a public health perspective, fat explosion is a key feature of modern life that will determine the health of a population, and where non-medical and medical measures need to be instituted.

3.1.3 Age Explosion

As the population explodes, the life expectancy grows. Until mid 1800s, life expectancy was averaging about 40-45 years. From then on, life expectancy in many parts of the world rose to about 75 to 85 years. This in itself is a sign of health. Simultaneously, it's also indicative of an aging population on the rise, which increases the demands on the health care system dramatically, as well on access to long terms funds (pension schemes, be they private or society-funded). For an aging population, health care may need to be organised differently compared to a younger and healthier population, in order to be financially sustainable.

3.1.4 Travel Explosion Pre- and Post- COVID-19

The global travel has been increasing over the last century. Just by looking at international tourist arrivals in popular travel destinations like Thailand, one can observe an increase of 4-5 times more travelers today compared to 1990. The ultra-long haul flights from Singapore or Australia to cities in the USA are late additions to the menu. Apart from the risk of thrombosis, the long-haul travel carry little was also a risk. Until COVID-19 hit.

The pandemic illustrated like never before how rapidly an epidemic that started in Wuhan in China could travel and germinate around the world, giving rise to mass infection and death on a scale that few people could comprehend. With modern DNA and RNA sequencing tool, the paths of spreading have to some extent been mapped, and the A, B and C variants and their entry to different regions are to some extent understood. International air travel can be seen as huge incubators, as can cruise ships. The post-COVID-19 situation will be quite different, given that there will be reticence to fly and that airlines will need to give more space to passengers. Obviously, if there would be a system to understand in real-time the emergence of new viruses or micro-epidemics of old ones like polio and measles, the situation could change.

3.1.5 Climate Change
This is one of the hottest topics (no pun intended) of today. Looking back over the history of the earth, this change in temperatures has remained a constant phenomenon. Global warming has happened before and it has subsided as well. The current debate about a possible man-made climate crisis is different from the debate in the 1950s when people feared a new ice age. By Googling 'new ice age', one easily finds data suggesting that a new ice age is in the making. Who knows? Maybe it is. However, with global warming, if it will be in the order of magnitude some models argue that it will have a list of health effects, including effects on access to water,

and nutrients, and this will steer where people can live and survive. In the catastrophic scenarios, man will no longer be here. But, the Universe is in no hurry and there are billions of new sites where life can start.

3.1.6 Global Middle-Classification and Changing Professional Roles

This may not yet be seen as a trend, but in large populations moving from poverty, this may be a change of great importance. When the global population moves away from severe poverty, the pattern of consumption and lifestyle changes and the need for and the ability to pay for health care increases. Although middle class in India is less than 4% of the entire Indian population now, it can be predicted that Indian middle class in 20 years' time will be the largest middle-class population globally. Their consumption patterns and food patterns are very likely to change, as well as demands for clean air, clean water and size and standard of living spaces. The spending power of the middle class will be a force to reckon with. The demand for qualified health care will rise in conjunction.

3.1.7 Sedentification

We move less and less, and this is a global trend. The middle classification and change of professional roles will enhance the development without risking mobility. Basically, with the food apps and remote working available nowadays, we never need to get far away from the sofa. The sedentification will have tremendous effects of health and mortality, linked to obesity, diabetes, cardiovascular disease and musculoskeletal disease. Since this is a universal phenomenon, we need to think of medical and non-medical methods to counteract this development.

The sedentification changes the cardiovascular system, the body fat distribution and the body shape. The combined effects of overnourishment, undermobility, sedentification and in some cases, increases in alcohol intake are difficult to foresee, but one could speculate that these combined factors would result in premature aging and earlier development of cardiovascular disease, cancer and musculoskeletal disease including mobility problems. Given that we are seeing mass changes in the way we live, there will be mass consequences

3.1.8 Pornification

Much of what we experience during life emanates from screen-based impressions and via the smart phones communication systems. The view of reality that we are being served is much different compared to what reached people's minds during the last century. The world as depicted in music videos, and in movies, as well as through internet access to very explicit and outré pornographic movie content creates internalised mind settings that are quite different from most in the previous decades. These are accessible to children from early ages even before normal concepts about procreation and sex would be formed. These evolving programming and mind sets will change behaviour, and behaviour changes will result in changed disease patterns. As was seen with the AIDS crisis, changed ideals, and changed behaviours caused a world epidemic, costing many young individuals their life. It put a lot of strain on the health care systems. Novel sexual habits, moving from conventional vanilla sex, changes infection patterns like HPV infection and leads to increases in oral cancers. All consequences are not mapped, but apart from obvious changes in infection patterns, there may also be effects on young people's

11

ability to form relationships resulting in children and perhaps even the rate of rape and domestic violence. Irrespective of what we believe in, there will be consequences of the ongoing pornification.

3.1.9 Reduction in Fertility and Fecundity

Fertility is the ability to create offspring. One key measure is number of children per adult female during her lifetime. Fertility in the western world is going down and is now in many countries at a level where the population cannot be sustained numerically. Italy is one of the most extreme countries, where fertility is down from 1.2 to 0.8. Apart from the demographic consequences, where a shrinking youth base and working population will need to shoulder the financial burdens of the society of a growing cohort of post-retirement, also the forming of personality and sentiments of connection or isolation may be affected. Maybe this will result in more psychosocially based morbidity in adult life?

Fecundity is the likelihood of childrearing to result in a fetus. Here of course, we talk about the basic biological mechanisms and probabilities of egg production and quality, the sperm numbers and quality, factoring in the tubas and the uterus and so on. Going forward, the techniques of in vitro fertilization and surrogate motherhood will continue to evolve, but all of these are costly and cumbersome.

3.1.10 Reduction in Birth-Related Death

In most western countries, with the exception of the USA, the rate of perinatal death of mothers and children continue to decrease, owing to better maternal care, and dedicated prenatal and neonatal intervention capability. In addition, the capability of caring for preterm born children is increasing. Preterm born children may at higher frequency display CNS defects and ocular defects that may result in morbidity prenatally and later would need health care. The numbers are small but depending on degree of injury the care of these individuals may be noticeable. So, should we abstain from salvaging preterm babies? Probably not, but there is a maturity limit to where the risk for future invalidity increases, and this risk cannot readily be mitigated.

Given that childbirth numbers are decreasing in the western world, we need to take very good care of the mothers and the offspring.

3.1.11 Food Intake Changes

Man is obsessed with food. There is an ongoing explosion in access to food, food innovation and spreading of ethnically defined types of food, like French, Italian, Spanish, Japanese, Thai, Indian, Chinese, Mexican, Brazilian and American and more. Even the food of smaller populations like those of Korea, Taiwan, Argentina and Peru have their food culture permeating across the globe, adding to choice and to quality of life.

The increased diversity is to some extent contrasted by the ongoing narrowing of food choice that groups of people choose in terms of vegetarianism, veganism, pescetarianism, fruitarianism and so on. For many, these food choices are temporary and for others these become lifelong habits, sometimes orderly followed and sometimes with fanatic obsession. Medically, it is clear that our digestive system is not built for a fully vegan diet, since we do not have the enzymes for production of some of the necessary nutrients like vitamin B-12 that meat provides. For babies and small children, extreme diets can have devastating effects, and for adults, narrow food choices may result in deficit-based morbidity. Many of us are experiencing an expanding non-meat culture and food choices, whose health consequences are not understood. Maybe this could result in lower incidence of bowel cancer, but at the risk of increase in vitamin-deficiency-based morbidity? Something to think about. The move away from dairy-based products may also have positive and negative health consequences.

At the same time, the supply chain of food is changing – with the two most obvious examples found in corn and palm oil. The food chain starting in the production of corn and then corn-fed animals and products containing high fructose corn syrup is expanding and may have health consequences. Similarly, the increasing dependency on palm oil for food production may have both health and environmental consequences.

3.1.12 Car Transportation and Car Access Explosion

This may sound like no news, but for large portions of the world population, owning a car is one of the most life-changing events. In the US and Europe, we have adapted to the system of individual car access and travel and to the sometimes devastating health consequences we face in traffic. Traffic-related invalidity increased dramatically during the last century, which is a mathematical function of a number of factors such as driving skills and experience, traffic rules and safety systems, car density, road quality, car passive and active safety systems, and navigation support systems.

As much as the cars are offering fantastic individual freedom, we humans are merely educated monkeys with limited attention span and limited understanding of reality. To allow such creatures to drive 1+ ton vehicles at high speed may eventually turn out to be not a very smart idea. Therefore, the ongoing quest for self-driving cars and added active safety will continue to help in reducing mortality, morbidity and invalidity. For nations undergoing car access explosion, there will be a steep learning curve and the consequences will be felt in many society systems, including health care.

3.2 The Best of Times

3.2.1 Biological Knowledge Explosion Translation to Medicines

We live in the best of eras when it comes to our ability to diagnose and treat disease. We are making progress on all fronts and our understanding of biology is increasing month after month. The combined explosions of DNA, RNA, protein, metabolism, cell structure and organ function knowledge and studyability make an unstoppable force that will bring us further towards our implicitly agreed goal of a life without disease. Think back 100 years and consider how much we knew; then imagine the speed of knowledge expansion that transpired. Then think of our knowledge today and the speed of evolution. It is mind-blowing that we now can create synthetic organisms, mini organs, edit DNA and to some extent read minds. Sobering though, the outlook for patients with severe organ disease is still poor as is that for patients with metastatic cancer, or severe autoimmune disease. In such fronts, much remains to be done.

Although our diagnostic capabilities in terms of imaging and lab tests are increasing dramatically the evolution of new therapies is slow. The paradigm of reductionism,

15

using animals for predicting effects in man has helped to some extent, but it has also introduced vastly incorrect signals. Both the difference in size, cell types, immune system, microbiome, food, and difference in disease spectrum are sources of this faulty predictability. In addition, many animal models have been constructed based on a certain idea of the etiology of the disease. Then of course, the therapy testing in this model will test whether or not the therapy will be active in this type of a biased system. However, if that paradigm and basic etiological idea is incorrect, then the a priori likelihood of an effect in man is zero. A typical example is Alzheimer's disease. Mice do not get Alzheimer's, so in order to study it in animals, the researchers have constructed animal models showing features of Alzheimer's based on concepts like neuron tangles and beta pleated sheet and accumulation of amyloid protein. In animals, a series of therapies have shown efficacy while they spectacularly failed in man. Classical examples of circular proof.

Disregarding these failures, the intent and activity is ongoing to create more humanised systems, better mimicking human conditions, and using cell systems from patients with the disease under study.

This chapter aims at providing a snapshot in time of the ever-ongoing progress of health technologies, with the explicit understanding that so much is happening at the same time and that no one can have the full picture.

16

3.2.2 Diagnostics Revolution

3.2.2.1 Imaging Explosion

3.2.2.1.1 Capture Systems

3.2.2.1.1.1 Ultrasound and Doppler Ultrasound

Medical ultrasound was first tested in 1953 at Lund University by a cardiologist named Inge Edler and Gustav Ludwig Hertz's son Carl Hellmuth Hertz to non-invasively depict heart function. This is a method that uses high frequency sound waves and has been tremendously developed over the years to be applied in multiple diagnostic and therapeutic medical areas. It is considered to be very safe. With the addition of computerised signal processing and AI, the resolution may be enhanced and further classification of tissue anatomy and consistency (e.g. like liver fibrosis) can be achieved. With the use of Doppler technology, blood flow parameters can be added.

3.2.2.1.1.2 Computerised Tomography

The 1979 Nobel Prize in Physiology or Medicine was awarded jointly to Allan M. Cormack and Godfrey N. Hounsfield "for the development of computer assisted tomography."[2] The practical invention was accomplished by Hounsfield in 1972 and user technologies and mathematics from a lineage of similar thinking by Cormack in the USA. By use of X-rays from multiple angles, a 2D and 3D image may be constructed, allowing precise diagnosis of a multitude of diseases. Further additions like single photon detectors, and computer-based methods to enhance image quality makes the method enormously valuable clinically. CT may be

---

[2]The Nobel Prize in Physiology or medicine 1979. (2021). Retrieved from https://www.nobelprize.org/prizes/medicine/1979/summary/

applied in most organs and can depict volumes down to less than 1 mm in side. With the addition of enhanced reality and 3D viewing, doctors may look into the human body and understand for example how to best perform surgery.

3.2.2.1.1.3 Magnetic Resonance Imaging (MRI)

Mansfield and Lauterbur were awarded the 2003 Nobel Prize in Physiology or Medicine for their "discoveries concerning magnetic resonance imaging."[3] The first MRI images were published in 1973. MRI uses the interaction of strong magnetic fields and body tissue to create images similar to computerised tomography, but without the x-ray component. Images add information that is essential in several medical diagnostic fields and show for example better depiction of intracerebral processes.

3.2.2.1.1.4 Positron Emission Tomography

No Nobel Prize has been awarded the PET technology directly, but the 1943 Nobel prize was awarded to Hevesy for the use of isotope tracers. The PET technology was developed in the USA and can use different tracers for different purposes. In cancer medicine, it is employed to show cancer metastases.

3.2.2.1.2   Image Management Systems
3.2.2.1.2.1 Sending, Receiving and Storing Images

Storage technologies both in terms of memory capacity and image compression techniques have enabled effective storage and retrieval of x-ray images and tomography-based imagery. Sending images over the

---

[3]The Nobel Prize in Physiology or Medicine 2003. (2021). Retrieved from https://www.nobelprize.org/prizes/medicine/2003/summary/

internet enables remote examination and allows expert diagnostics from afar.

3.2.2.1.2.2 Virtual Reality

3D technologies and enhancement techniques have become generally available during the last few years, although the philosophical concepts and imaginary existence have been described since long before. The last few years have seen multiple examples of the use of virtual reality and enhanced reality as well as 3D imaging in health care. These technologies increase precision of radiation therapy and of complex surgery, for example in the head and neck region or intra-cerebrally.

3.2.2.1.2.3 Artificial Intelligence

Nowadays, AI is increasingly being used in image analysis and in clinical diagnosis. Typical examples are uses in screening technologies like mammography or lung cancer screening, where a machine-based non-biased methodology tend to yield better precision. Additionally, artificial intelligence is being used in selection of drug targets in designing molecules that can inhibit or stimulate these drug targets. This is just the beginning, and although AI has been in the making since the 1950s, much more is yet to come in this technology. Definition of patient categories which may have shared disease mechanisms and the understanding of which therapies should be tried in which patients are some of the key areas where AI can help.

3.2.2.2 *Biomarker Explosion*

3.2.2.2.1 Humoral Biomarkers

A blood sample is one of the simplest forms of not so invasive diagnostics. Although the blood is a veritable pea-soup of corpuscular (red blood cells, white blood cells and thrombocytes) and non-corpuscular elements (sugars, proteins, lipids, metal ions, non-metal ions, DNA, RNA etc.), it is quite hard to understand how anything can be discerned. However, the art and science of clinical chemistry has evolved during the last 100 years and bioanalysis can now pick out and detect minute amounts of substances in the blood. This means that early signs of organ injury can be shown (like early signs of myocardial infarction or signs of liver cell injury).

Blood glucose is one of the most typical biomarkers and is used to understand the need for, adjustment of, and adherence to antidiabetic therapy. Another example is something called HBa1C, which has been shown to correlate with long term diabetic disease control.

The immune system is extremely complex but both antibodies and lymphatic cells with certain reactivity can be detected to diagnose for example autoimmune disease where the immune system reacts with antibodies and/or tissue destroying immune cells against healthy tissue.

Another area of development is the area of liquid biopsies. The idea is that by using blood samples, one can get a picture of what is going on in a tumour or in an organ. One approach is to analyse thrombocytes which pick up snippets of DNA and RNA when passing through tumours. Another method is to detect circulating tumour DNA. This provides a measure of remaining tumour after surgery. Another aspect is to determine which are the driver mutations at play (genome changes which promote tumour growth), and to adjust therapy on that basis. Circulating tumour cells can provide similar information and may be used as a complement to cancer tissue biopsy.

The methods to detect small amounts of anything in the blood or in other body fluids are continually improving and will lead to increased precision and greater granularity of diagnosis of both acute and chronic disease, as well as determination of the most suitable therapy.

3.2.2.2.2 Olfactory Biomarkers

This is an interesting field. Dogs have been trained to detect lung cancer, and it was recently reported that Parkinson patients have a typical acrid smell. Artificial noses have been developed and are now being trained to detect a number of diseases. Changes in metabolism of sugars or fats or proteins may give rise to substances that can be detected in exhaled air.

Another example of exhaled air detection is the measurement of exhaled nitric oxide, which is already used to guide asthma therapy.

We can assume that these technologies become increasingly precise and sensitive and that diseases which formerly only could be detected at a fairly late stage could, by using the combination of clinical signs and symptoms and clinical chemistry and radiology, in the future help detect it at an early stage in the exhaled air.

3.2.3 Therapeutic Explosion

3.2.3.1 Medicines Explosion

3.2.3.1.1 Small Molecules

The first small molecule medicines (<900 Daltons) were created in the late 1800s. Paracetamol was invented after a mistake in the pharmacy had a patient with fever receive acetanilide instead of naphthalene as a remedy for a worm infection, and the fever was reduced. A one-patient experiment that soon led to a widespread clinical use of paracetamol. Medical chemistry was in its early stages and only fairly simple compounds could be synthesised. The path from synthesis to clinical testing was seen as very short. The growth of the pharmaceutical industry was to a great extent built on developing small molecules. Areas like anti-hypertensives, anti-diabetics, anti-inflammatory, anti-psychotic, anti-depressant, hypnotics, anti-asthmatic compounds are historically small molecules and many new compounds – for example in the cancer area where personalised medicine focused on receptors that are involved in driving the cancer – still are tempted to be blocked using such small molecules. These bind to the receptor and block its activation and therefore put a break on the cancer, adding months and years to the life of the patient.

Also, the anti-diabetic area has seen new medicines arrive which are small molecules and are based on new insights into how insulin is secreted or how glucose is being reabsorbed in the kidneys.

In neurology, small molecules also play a big role but getting a drug to the right place in the brain at the right concentration, blocking the right receptor in the right cell is quite a difficult task and there are fewer successful examples.

3.2.3.1.2 Large Molecules

Large molecules are often antibodies, which are longer strings of amino acids (polypeptides). This is a fairly late innovation (Milstein and Köhler 1975) and links back to technologies that use viruses to immortalise mammalian cells and technologies where antibody producing cells are artificially fused with antibody producing blood cancer cells, to create a machinery that keeps producing antibodies of a defined kind – monoclonal antibodies.

By creating antibodies that bind to a specific receptor on the cell surface or to a molecule that is creating the disease, the receptor signaling can be inhibited and in the case of an overabundant bioactive molecule, its concentration can be lowered. In that way the disease activity can be reduced, and the symptoms can be diminished and organ damage can be limited.

Large molecules have been among the most successful drugs during the last 20 years, and the latest number one is called Humira (Human Monoclonal Antibody in RA). It is a fully human antibody that targets a bioactive molecule called Tumour Necrosis Factor alpha which is overabundant in rheumatoid arthritis. In 2018, this drug sold for 19.8 billion US dollar. This type of medicine has meant a huge difference for patients with RA who historically have been using quite toxic medicines, which have not been able to control the disease. With these new medicines, there is less destruction of joint cartilage and many more patients can avoid invalidity.

Also, in cancer therapy too monoclonal antibodies are being employed. One example is so called anti-programmed cell death receptor 1 (anti-PD1), which has brought a paradigm change in the treatment of cancer. The previous therapy paradigms included surgery, radiation, targeted medicines (molecules intended to put a break on cancer drivers) and now changing into immuno-oncology. This is the insight that cancer uses a number of methods to avoid or to stop immune attack and by blocking these methods we allow the immune defense to effectively recognise and kill cancer cells. The leading anti-PD1 is expected to yield approximately 20 billion US dollars in sales in 2015. The paradigm shift has totally changed the way we think about cancer therapy and has set forth an unprecedented number of new drugs in development.

3.2.3.1.3 Peptides

Peptides have been used as medicines since the 1980s. One of the best know is desmopressin (DDAVP) which is used as an anti-diuretic for nocturia, and as a second indication von Willebrand's disease or mild hemophilia A. The list of peptide drugs has been growing.

3.2.3.1.4 RNA Based

2018 saw the approval of the first mRNA based therapeutic. Patisiran, a medicine given as a lipid nanoparticle containing an siRNA (small interfering ribonucleic acid) which inhibits the production of an abnormal form of transthyretin (TTR) which role is to transport vitamin A in the body. With very few injections, this medicine reduces the production and subsequent deposition of the TTR protein in tissues and improves a clinical variable measuring the effects of polyneuropathy on the patients' abilities. The placebo treated patients worsen though.

There are many other forms of RNA based therapies in clinical research. For example, mRNA (messenger ribonucleic acid) can be used in vaccines against infectious disease or cancer vaccines to direct the immune system to attack and kill cancer cells. We now see the first examples of mRNA-based vaccines in the BioNtech and Moderna vaccines approved for COVID-19. The most beautiful feature of the vaccines is that the antigenic portion, which is a part of a protein or a peptide, can be shown to antigen presenting cells like the dendritic cells, without having been visible in the blood. The mRNA strands coding for these antigenic structures are taken up by the antigen presenting cells and are then expressed so that immune cells are instructed to either form antibodies against the antigen or to muster a cellular response against these structures. That is the basis of the vaccine. The immune defense can then attach the virus as though the immune system would have experienced a previous infection. The mRNA used is out of the systemic circulation in a few hours compared to other types of vaccines, which can linger for days, weeks and months. With the mRNA-based vaccine, a reconstruction is then a speedy affair if the virus mutates and is using other structures to enter and injure cells.

3.2.3.1.5 Gene Therapies

Many devastating diseases are caused by abnormal and deleterious genetic diseases. This could be genetic defects leading to lack of formation of a certain protein or to production of abnormal molecular forms of proteins, which in turn will have biologically bad consequences. The genetic background has been understood for quite some time, but it has been and still is technically difficult to get the correct genes into the cells that are key for the disease manifestations. Systems to achieve this goal using viral constructs as a way to direct the genes to the right cells have been tried with very limited success.

In 2019, we saw the start of the first CRISPR based human study. It is believed that this technology is a way to reprogram specific genes using the CRISPR enzyme and RNA sequences, which guide the enzyme to the right DNA sequence place. The first patients have cancer and further studies are under way to treat blood disease like sickle cell disease b=and beta-thalassemia. Also, a form of congenital blindness may be investigated (Leber congenital amaurosis – a gene therapy was approved already in 2017 for one type of defect, and the current approach targets another abnormality). All is not known about the short and long term effects of these interventions, and for example CRISPR which was first thought to be causing very specific genetic alterations has been shown to also give rise to off target mutations.

In May 2019, the FDA approved a gene therapy for spinal muscular atrophy. Children with the disease die early because of muscle weakness.

There are 6000 genetically based diseases where these developing technologies may eventually be employed, and the majority of the diseases can theoretically be corrected with genetically targeted therapies.

3.2.3.2   Cell Therapies
3.2.3.2.1   Cell Therapy: A New Therapeutic Modality
Over the last few decades, we have seen the main therapeutic modalities for human disease transition from small molecule compounds made via chemical synthesis to large molecule compounds such as monoclonal antibodies made via cell factories in bioreactors. Over the last decade, a new modality has emerged on the therapeutics landscape: cell therapy. It uses the cells themselves as a therapeutic agent and infuses them into the body, rather than the cellular products derived from cells.

3.2.3.2.2   Cell Direct and Indirect Effects
The therapeutic effects exerted by the cells in cell therapy can either be via cellular direct effects, whereby the cells

which are infused into the body are the direct effector agents, that either target malignant cells, such is the case in cancer, or replace certain dysfunctional cells such as pancreatic beta cells, as is the case in Type 1 diabetes. On the other hand, there are also examples where the cells are not the direct effector agents. Rather the cells are factories infused into the patient's body which produce biological factors that elicit the biological function via paracrine effects. An example of this would be mesenchymal stem cells which are infused into a patient's body and the observed clinical benefits are attributed not to integration of the cellular products, but rather due to factors such as cytokines produced which broadly dampens an inflammatory response.

3.2.3.2.3 Ex Vivo and In Vivo Gene Delivery

Cell therapy can sometimes be characterised as *ex vivo* gene delivery, as in some instances of cell therapy, the infused cell product has its genetic material altered using viral methods (predominantly lentiviral vectors). An example of this would be chimeric antigen receptor T cells such as Kymriah and Yescarta which have been approved for treatment of hematological malignancies. Another modality that has steadily been gaining prominence is *in vivo* gene therapy, which is the direct infusion of genetic elements into the human body. This typically happens via packaging into viral particles, and the most commonly used viral vector would be Adeno-associated viruses, AAVs. In addition to AAVs, there are multiple efforts underway to explore non-viral vectors for gene delivery, such as lipid nanoparticles.

3.2.3.2.4 Ex Vivo Gene Delivery: CAR-T Cell Therapies

In the realm of cell therapies, the path-finder therapies have been CAR-T cell therapies, namely Tisagenlecleucel marketed by Novartis for the treatment of acute lymphoblastic leukemia and relapsed or refractory diffuse large B-cell lymphoma, as well as, Axicabtagene ciloleucel marketed as Yescarta by Kite Pharma for the treatment of Non-Hodgkin Lymphoma and relapsed or refractory diffuse large B-cell lymphoma. Both these therapies involve removal of T cells from the cancer patients, engineering the T cells using lentiviral vectors to target CD19 receptors on the B cell surface, and re-infusing the cells back into the patient. The cost for the treatment is in the ballpark of US $400,000 in the United States.

Beyond CD19, multiple biopharmaceutical companies are exploring development of cellular therapies targeting other surface proteins beyond CD19, on the cancerous cells, to broaden the target indication beyond B cell malignancies. It has also been reported in patients treated with CD19 CAR-T cell therapies, that tumour cells may start to lose CD19 expression, or outgrowth of CD19negative tumour cells start to develop. Hence, it would be important to develop new cell therapies with targets beyond CD19. Frequently pursued targets beyond CD19 in CAR-T therapies in the hematological setting include BCMA, CD22, CD30, CD20, CD123 and CD33.[4]

---

[4] Leick, M. B.; Maus, M. V. (2019). CAR-T cells beyond Cd19, Uncar-ted territory. American Journal of Hematology, 94(S1). doi:10.1002/ajh.25398

To increase the clinical efficacy of the newly approved CAR-T therapies in various hematological malignancies, companies are also exploring various combinations of CAR-T with therapeutic antibodies such as, immune checkpoint inhibitors PD-1/PD-L1, cancer vaccines and oncolytic viruses. The premise of such combinations is that immune checkpoint inhibitors can help activate an immune system which may be experiencing immune exhaustion. An example of such a combination is the combination of Yescarta with Pfizer's mAb utomilumab which is an investigational 4-1BB agonist. Preclinical evidence suggests that 4-1BB is upregulated upon exposure to CD19-expressing tumour cells and this would help further increase T cell anti-tumour activity.

3.2.3.2.5 CAR-T Cell Therapy: Moving Beyond Liquid Tumours

Cell therapy has its beginnings in hematologic malignancies, many groups are exploring the possibility of extending CAR-T therapy to the solid tumour setting. There may be several issues with replicating the success of CAR-T therapy, one being the difficulty of infused T cells to easily access the tumour. Many solid tumours have very immunosuppressive microenvironments around them which would render the infused T cells ineffective. Another potential reason that the CAR-T therapy works well in hematological malignancies is that the dispersed nature of the tumour allows for continued antigen stimulation of the CAR-T cell population, allowing for sustained activation.[5]

---

[5] Junghans R. P. (2017). The challenges of solid tumor for designer CAR-T therapies: a 25-year perspective. Cancer gene therapy, 24(3), 89–99. https://doi.org/10.1038/cgt.2016.82

3.2.3.2.6 Solid Tumours: TCRs

In addition to CAR T cells which are engineered to recognise proteins that are recognised on the surface of the cells, companies like Juno Therapeutics have also made a foray to develop T cells expressing engineered T cell Receptors (TCRs). The TCRs are engineered through optimization pf TCR-alpha and beta transgene pairing to recognise intracellular tumour specific proteins that are presented by the major histocompatibility complex (MHC) on the surface of the tumour cell. Beyond gene transduction to generate CAR-T cells or TCR therapy, companies such as Tessa Therapeutics have also utilised non-gene transduction approaches, whereby they take autologous T cells from patients and grow them in conditions to select for T cells that are primed against the tumour cells, for example activation of the Epstain Barr Virus-specific T cells for activity against EBV related tumours such as Nasopharyngeal Cancer.

3.2.3.2.7 Cell Therapy: Beyond T Cells

Beyond T cells, several groups have also started to look into other cell types such as NK cells, which may have similar anti-tumour effects without side effects associated with T cell therapies such as Cytokine Release Syndrome or Graft versus Host disease. NK cells may also offer the opportunity for allogeneic cell therapy, as there is no need for HLA matching between donor and host cells. This holds the promise for an off-the-shelf product with greatly depressed CMC costs and eventual therapy costs to the patient.

3.2.3.2.8 Cell Therapy – Cell Replacement

Another manner in which transplanted cells can be the direct effector cells, is in the context of cell replacement. In diseases such as Type 1 Diabetes, the disease occurs as a result of the failure of one cell type to perform its function, in this case, it would be failure of pancreatic beta cells to modulate glucose metabolism. Since the first pancreas transplantation in 1966, the process has been much improved due to surgical advances, as well as improved conditioning protocols and donor selection. However, one big challenge in the field of pancreatic transplantation, remains a steady and scalable cell source. The advent of new cell biology techniques, such as the generation of IPS cells and subsequent differentiation protocols to derive a well characterised and homogenous beta cell population, holds promise to address these bottlenecks. Moving forward, it would be important for the cell purity issues to be adequately address in view of potential teratogenicity, before such technology can realise its promise and be scaled and widely implemented in the clinic.

3.2.3.2.9 In Vivo Gene Delivery

The past decade has also seen the advent of *in vivo* gene therapy, as described earlier in this chapter. In 2017, Luxturna (voretigene neparvovec) was the first gene therapy approved by the FDA for a hereditary disease, in this case for the treatment of individuals with inherited retinal dystrophy due to biallelic mutations in the RPE65 gene. Patients would also need to have viable retinal cells remaining. It is a prescription gene therapy product using AAV technology to administer the normal RPE65 gene to the eye. Hence, it serves to replace function but not through cell infusion, rather it works through genetic modulation of existing cells within the patient. The therapy is priced at US $425K per eye and has also sparked discussions around payment models built around patient outcomes. The gene therapy innovator company

31

would only receive payment if the patient reports positive clinical outcomes.

The second gene therapy approval for a hereditary disease came in 2019 for Zolgensma, for the treatment of spinal muscular atrophy, an autosomal recessive neuromuscular disease associated with biallelic mutations in the survival motor neuron 1 (SMN1) gene. Zolgensma uses an AAV vector delivered either intravenously or intrathecally into the cerebrospinal fluid, which delivers a copy of the SMN1 gene which makes the SMN protein, that is required for the maintenance of survival of motor neurons. Zolgensma allows for motor neurons to make the requisite amounts of SMN protein enabling neuron survival and function.

These two recent approvals would hopefully pave the way for the development of many other gene therapy products. One current bottleneck in the development of gene therapy candidates is the global shortage of clinical scale GMP grade AAV manufacturing facilities. In future, one would also expect a shortage of commercial scale AAV manufacturing capacity. The AAV manufacturing process is still very much laboratory scale, scaled-out, rather than an industrialised process that has been scaled up. The AAV manufacturing process is very manual and presents opportunities for further optimization and industrialization. The inefficiencies in AAV manufacturing has made it difficult to generate large numbers of viral vectors for systemic administration and this has hampered the expansion of disease indications that can be potentially treated with gene therapy. Currently most of the gene therapy trials seek disease indications that allow for more localised delivery, for example delivery to the eye or to the spinal cord.

For gene therapies delivered via AAV vectors, the patient must also be tested for immunogenicity (anti-AAV antibodies) prior to treatment. Limitations due to prior AAV exposure, the high cost of AAV manufacturing, as well as, recent concerns related to potential integration of AAV vectors into the genome, have also paved the way for development of other viral delivery methods. The most mature alternative would be lipid nanoparticle formulation to encapsulate the genetic elements. Many groups globally are working to optimise and improve lipid nanoparticle formulation, accelerated by the development of optimised ionizable cationic lipids, to reduce immunogenicity and to increase targeting to non-hepatic tissues. The promise of lipid nanoparticle formulation is the ability to deliver larger payloads, the ease of manufacture, as well as, the possibility of repeated dosing.[6]

3.2.3.2.10 N=1 Therapies
In addition to the recent flurry of activity by biotech companies in development of gene therapies, the last few years have also seen an increased interest in N=1 therapies that are usually orchestrated by academic medical centers. A classic example of this is Milasen, which was a personalised antisense oligonucleotide (ASO) developed for a splice defect related to Batten's disease for a 6-year-old girl which restored function of the MFDS8 gene. The entire process of sequencing the genome, identification of the mutation, generating candidate ASOs, doing abbreviated toxicology studies,

---

[6] Kulkarni, J. A., Cullis, P. R., & van der Meel, R. (2018). Lipid Nanoparticles Enabling Gene Therapies: From Concepts to Clinical Utility. Nucleic acid therapeutics, 28(3), 146–157. https://doi.org/10.1089/nat.2018.0721

administering the treatment took slightly more than a year.[7] This was an impressive, cross disciplinary effort that united in their singular sense of purpose to ameliorate the symptoms of a little girl. As gene sequencing becomes more accessible, gene therapy design capabilities grows more prevalent, we may start to see many more such personalised gene therapy efforts.

3.2.3.2.11 Increasing Access to Cell Therapy and Gene Therapy Products

Even as cell and gene therapy products hold tremendous promise, increasing accessibility to the promised clinical benefits is hampered by the high cost of goods, particularly around the high manufacturing and logistics costs. One way in which the cost could be substantially reduced is through a transition from autologous cell therapy products to allogeneic cell therapy products. The cell therapy products approved currently require cell sources that are derived from the patients. This greatly complicates the manufacturing supply chain, as the patient needs to undergo apheresis for removal of their peripheral blood mononuclear cell populations. The cells need to be shipped to a central manufacturing facility where they are modified according to protocols, and if needed, genetically modified. The cells then need to be

---

[7] Kim, J., Hu, C., Moufawad El Achkar, C., Black, L. E., Douville, J., Larson, A., Pendergast, M. K., Goldkind, S. F., Lee, E. A., Kuniholm, A., Soucy, A., Vaze, J., Belur, N. R., Fredriksen, K., Stojkovska, I., Tsytsykova, A., Armant, M., DiDonato, R. L., Choi, J., Cornelissen, L., … Yu, T. W. (2019). Patient-Customized Oligonucleotide Therapy for a Rare Genetic Disease. *The New England journal of medicine, 381*(17), 1644–1652. https://doi.org/10.1056/NEJMoa1813279

shipped, post-processing, back to the patient and re-infused by medical professionals.

Allogeneic cell therapy product promises to deal with the front end of the manufacturing process. Rather than having to source cell products from the patients and having to deal with issues around low viability of cells, as well as, variability in the transduction efficiency. If the cell products can be sourced from a uniform source and easily scaled up, it would greatly simplify the logistics of the cell therapy procedure. In addition, it would also allow for more predictable cell dynamics during processing as the cells would have all been derived from well phenotyped and banked source.

Additional technological advances that would reduce the manufacturing cost cell and gene therapy products also lie in analytics. Take for example AAV manufacturing. Currently the process is a very labourious and low efficiency process. Traditionally, triple vector transfection of adherent cell lines is used, and this is basically a scaled-out process, rather than a scaled-up process. In order for the efficiencies to be realised through batched manufacturing using a bioreactor setup similar to mAb manufacturing, the process needs to be transformed into one utilizing adherent cell lines. In addition, if stable producer cell lines could be derived, it would circumvent the low efficiency associated with transient transfection. Lastly, analytics would also need to be correspondingly upgraded. Currently it is challenging to purify the resulting AAV batches, to separate the AAV product from empty capsids. Novel chromatographic methods employed for these applications hold promise to greatly improve high purity and high titer AAV manufacture.

3.2.3.2.12 Cell Therapy – Indirect Effects – Unproven Promise

In addition to the cell direct effects outlined by the various cell and gene therapy products and candidates outlined earlier in this chapter, cell therapy products that elicit a clinical effect through paracrine effects have also started to proliferate. A commonly cited example of this would be mesenchymal stem cells are being tested for use to accelerate wound healing. This area of cell therapy is fraught with much controversy. In 2018, Dr. Anversa from Harvard Medical School and Brigham and Women's Hospital in Boston was found to have fabricated data for 31 publications. Some of these publications falsely claimed that stem cells were able to regenerate damaged cardiac muscle when transplanted into the heart; others claimed that resident cardiac stem cells were able to regenerate and replace damaged cardiac tissue. Multiple other research groups were unable to replicate these findings and subsequent investigations found that the primary data was falsified, setting back the field of cardiac tissue regeneration. Despite the retraction of these scientific papers, there remain clinics in the world which still make promises of cardiac regeneration with stem cell transplantation. These claims need to be carefully evaluated to ensure patient safety.

3.2.3.2.13 Cell Therapy at an Inflection

The field of cell therapy is one that holds much promise, the past decade heralded the arrival of this therapeutic modality into clinical mainstream with the approval of CAR-T therapies for treatment of hematological malignancies. Beyond ex vivo gene therapy, we also saw the approval of *in vivo* gene therapy assets for rare diseases. Hopefully, as technology matures, cell therapy and gene therapy products' efficacy can be extended to other solid tumours and wider disease indications. The adoption of these therapies would also be predicated on cost and scale, again technology in manufacturing, process development as well as novel cell sources hold the promise to allow this asset class to impact a larger patient population and yield larger clinical impact.

3.2.3.3 Surgery Evolution

3.2.3.3.1 Instruments and Anesthetic Techniques

Surgeons have used more or less sophisticated instruments since long – saw, tourniquet, knives, scissors, tweezers, sutures and on. Opening the abdomen or thorax or skull and removing diseased tissue or organs (e.g. appendix, gall bladder, kidney stones, cancer surgery, vascular surgery and endocrine surgery) have been developing through the 20th century. The surgical revolutions include the aseptic revolution, the anesthetic revolution, the X-ray revolution, the safe surgery revolution with very specific surgical curricula. Transplantation surgery revolution is linked to the immunology revolution. The vascular surgery revolution allowing autologous insertion of new vessels to replace calcified and obliterated coronary arteries is another. Hepatic and brain surgery have also been revolutionised during the 20th century.

Key-hole surgery which causes minimal tissue trauma has also been developed with specific surgical instruments with long arms allowing small size cutting, holding and sewing. Nowadays organs, which are hidden in the deep layers of the body like the kidneys or the prostate, can be reached by these technologies.

A further development is robotic surgery which may allow remote surgery so that patients do not need to travel to the surgical specialists.

3.2.3.3.2 In Surgery Sampling

3.2.3.3.2.1 Mass Spectrometry to Show Cancer Removal

New technologies are being developed to sample tissues as the procedure develops. For example, there is now being developed a surgical knife which can sense which tissue is normal and distinguish from various tissues such as cancer tissue, so that the surgeon can excise the cancer while sparing normal tissue.[8]

3.2.3.3.2.2 Prosthesis Revolution

The advent of 3D printing has created a new era for creating personalised prostheses. Increasingly these are also being functionalised so that they contain flexile power and sensing/feedback mechanism to fine tune grip and maneuverability.

3.2.4 Med Tech Explosion

3.2.4.1 3D Print

Computer assisted design and fast prototyping via 3D orienting technologies speeds up development of all kinds of medical devices such as inhalers, pacemakers, surgical instruments, and prostheses. Cells can also be used for 3D printing so that models of organs can be built using multiple cell types. This is the beginning of very

---

[8]Cheng, M. (2013, July 19). New surgical knife can instantly detect cancer. Retrieved from https://eu.usatoday.com/story/news/world/2013/07/17/surgical-knife-cancer-detection/2525877/

exciting development where the combination of stem cell technology and 3D printing may offer replacement parts.

3.2.4.2 Electroneural Interaction

We can now implant hearing aids and visual aids into CNS that allow visual and auditory signals to flow from extracranial adaptors to the central nervous system. Moreover, electroneural stimulation may create movement to paralysed limps and may even affect the immune system and provide non-drug benefit to patients with rheumatoid arthritis.

3.2.4.3 Pacemakers for the Stomach

Diabetic gastroparesis affects many patients with diabetes mellitus. Now it is possible to surgically insert a pacemaker device that stimulates the emptying of the stomach and normalise the function of the gut. This may be revolutionary for these patients.

3.2.4.4 Exoskeletons

Old age and frailty and falls are linked and in Japan where the number of centenarians are projected to rise from 70 000 in 2020 to 680 000 in 2050. According to the projections by the Japanese population research institute, the number of centenarians will reach 128,000 over the next decade; by the middle of the century, there will be more than 680,000.[9]

3.2.4.5 App Based and Soft Systems Revolution

It seems that soon, any medical problem will have its own application. Sleep problems, snoring, weight loss,

---

[9] McCurry, J. (2010, August 10). In graying Japan, scandal over "missing" 100-year-olds. The Christian Science Monitor. https://www.csmonitor.com/World/Asia-Pacific/2010/0810/In-graying-Japan-scandal-over-missing-100-year-olds

fertility, all can be searched, and you may find an app that purports to help. Also, online medical appointments and medical consultation can be reached via your smartphone and you may in the next quarter buy your prescription drug at the online pharmacy. This is a wave that moves very quickly.

3.2.5 IT/Data/Information/Big Data/AI/Deep Learning Explosion

3.2.6 Funding Explosion

3.2.6.1 Public Funding of Biological Research

Public funding of biomedical research has contributed massively to medical progress and new medicines. In the USA, institutions such as NIH, and NCI contribute huge amounts yearly and in EU the Horizon 2020 initiative has also been funding large scale medical research. These sums have grown during the last 20 years. The Horizon 2020 amassed 20 BUSD Euro over a 10-year period. A new program has been decided to the sum of 113 BEUR. The US yearly spending is in the order of 80 BUSD.

3.2.6.2 Venture Capital

Venture capital is vital to growth of the biopharma industry. The EU musters 15 BUSD yearly, which is dwarfed by the US 45 BUSD (Ernst and Young). This difference plays out in both the number of new potential medicines which may be researched, and also, importantly the speed over ground – i.e., how fast these projects progress in development. The increased access to venture capital is one of the key parameters in the growth of the number of biopharma start-ups and the rate of innovation in the industry. Thus, not only the quality of the innovations and the quality of the team progressing these ideas, but also the level of funding and the access to experienced company developers and board members determine the rate of successful exits and the rate of bringing new medicines to man.

3.2.6.3 *Stock Market Funding – Mega IPOs*

Typically, start-up companies go through funding cycles from early funding by family and friends and business angels, with growing rounds of funding – called series A, B, etc. These rounds of funding parallel the increasing detail in the project's description and the reduction in risk over time. A typical round A in EU is 15-30 MUSD and round B 25-45 MUSD. The corresponding figures for the USA are about double. Early biotech companies often go to the stock market to find capital, using the concept of initial public offering (IPO). The sizes of these offerings (a certain number of shares at a predetermined price) are intense affairs where many factors play together (company data, number of shares, share price, market conditions, and so on). The size of IPOs has risen over the last decade, and it is now not unusual to see IPO sizes over 200 MUSD. This capital inflow comes with reduction in ownership of the parties who invested initially in the company but gives more fuel for continued development of the projects of the company. Therefore, there is a higher likelihood of achieving important milestones and so-called value inflection points, i.e., Timepoints when new valuable data get presented, which alter the company's value.

In 2018, the top biotech IPOs raised 4.5 BUSD. Obviously, the biotech investment is speculative since the compounds are early phase and unproven, and the industry is ridden by 95% attrition, which is the percentage of all compounds entering phase 1, which for various reasons fail.

The allure is that the company hits gold and gets bought by one of the big pharma companies for large sums of money. In 2019, for example, Array Biopharma was bought by Pfizer for 11.4 BUSD since the company had received approval for two cancer drugs, and Eli Lilly bought Loxo Oncology for 8 BUSD, providing a windfall for the investors. In 2017, Gilead bought Kite Pharma for 12 BUSD. In 2018, Celgene agreed to pay 9 BUSD for Juno Therapeutics. These buyouts were clear wins for the initial investors and for the top officers of these companies, who typically receive substantial packages.

In spite of the downturn caused by the COVID-19 epidemic, the IPO circus has moved on and there seems to be no end in sight.

3.2.7 Personalised Medicine

Many of the existing medicines offer a low degree of efficacy, measured as the Numbers Needed to Treat (NNT). For example, statins have an NNT of 83 for saving a life (83 patients need to take statins for five years in order to save one life), and 39 for preventing a non-fatal heart attack. This is pretty unimpressive, and there is a call for higher precision of the therapeutic effect. The key is selecting patients, and in the case of statins, the simple measurement of plasma cholesterol may not be enough to discern those potentially benefitting from the drug. The opposite side of this measure is the Number Needed to Harm (NNH). In the case of stains, this is 10 for muscle damage and 50 for developing diabetes. Obviously, if the cost is low and the harm can be monitored, then this may be a good proposition. If not, one could think of better ways to spend money.

Based on these observations, there is a trend towards personalised medicine. This is best described as the right medicine at the right dose for the right patient. Effectively, the matching of the medicine to the patient needs to be much better than now and that the number of patients taking drugs but obtaining no benefit should radically decrease. The NNT should increase while reducing the NNH.

To do this, diseases need to be better defined and there needs to be the development of patient selection measures (often biomarkers, so called companion biomarkers that redeveloped in parallel with the medicine). These measures could also be phenotypic markers such as body characteristics like body weight or physiological measures such as cardiac ejection fraction (volume proportion ejected from the left chamber of the heart, which is low in heart failure). Patients with above a certain body weight or those with an ejection fraction below a certain number are more likely to benefit from the therapy.

3.3 Learnings from the COVID-19 Epidemic

The COVID-19 epidemic is and has been a real test on mankind's abilities on so many levels.

- The ability to foresee and prevent catastrophes like this
- The ability to quickly understand the start of the epidemic and to initiate countermeasures
- The ability to communicate and to warn other countries
- The ability to set forth a concerted and coordinated policy response

- The ability to set forth a concerted and coordinated scientific response
- The ability to perform full clinical disease understanding
- The ability to detect, track and isolate virus infection
- The ability to institute measures to confine virus infection geographically and to stop the spread
- The ability to use the best possible science and technologies to develop effective therapies, including clinical management, stopping of viral replication, quenching of cytokine storm, prophylaxis and management of thromboembolic events, antibody-based therapies, vaccine therapies, anti-inflammatory therapies, anti-fibrotic therapies
- The ability to lead and risk-manage drug development against the disease
- The ability to create rules for de-escalation of lockdown

And the list can go on. It is clear that we as a race have failed many of the challenges, and despite money pouring into COVID-19 related research, there is no prophylaxis and no cure yet.

We see vaccines being approved with the BioNtech vaccine being first and now a slew of other vaccines like Moderna, Sputnik V, the AstraZeneca vaccine, Sinovac and an Indian variant. These vaccines are being used in the biggest real-time experiment in the history of humankind. Whether this will be successful is still debatable, since the virus mutates, and we need over 70% of the population to be vaccinated before we can reach herd immunity. That is an expedition of enormous proportions and very much is at stake.

So, even with the level of sophistication that we have reached (and it is certainly not to criticise us, the human two legged monkey who has reached heights of sophistication that I think few would have guessed 200 years back) our thinking timelines, our action timelines and our technological advancement level, as well as our production timelines are such that the design, the research, the approval and the production of a vaccine for the full human population will likely take between 2-4 years.

The vaccination level will need to be in excess of 60%, although a deeper understanding of herd immunity is now being developed. Although the equation used to calculate the level of immunity needed to block outbursts of infection is simple and relies on the average number of person one infected individual spreads the disease too, there may be local or individual circumstance that allows a single person to accomplish mass spread like in confined spaces with lots of close contact.[10]

Another aspect is the identification of viral structures against which one would develop a vaccine. The COVID-19 virus particles are basically small fatty packages full of protein and RNA and use the viral spike on the surface of the particle to attach to and infect cells. These spikes are the most natural part of the virus to block, since blocking viral entry should stop the start of the infection. Many of the RNA and non-RNA-based vaccines are

[10]Hartnett, K. (2021). The tricky math of herd immunity FOR COVID-19. Retrieved from https://www.quantamagazine.org/the-tricky-math-of-covid-19-herd-immunity-20200630/

focusing on the spike proteins. Further, the process by which a certain individual's immune system will respond is not fully known, and since the virus seems to be able to elicit some kind of autoimmunity, the vaccine may in some individuals cause just that, since the immune system will develop antibodies and cell based immunity which may hit normal tissue and normal physiology. So, an individual does take a risk when agreeing to be vaccinated, and this risk needs to be weighed against the risk of i) becoming infected, ii) getting ill, iii) risk of long-term COVID-19 sequelae. It's not a simple choice for anyone.

Mass testing of the vaccine is not without its risks. With an accelerated scheme, you may cut corners that should not be cut. Once you have vaccinated a person, you may alter his or her immune system forever, and I think all would agree you cannot accelerate time. A one-year experience post-vaccination takes exactly one-year study to register data and longer to make meaningful conclusions.

Safety problems that only show up after a certain time period but in a low percentage of individuals may initially go undetected and may become a real problem when you start vaccinating billions of people. Trading a relatively low risk of bad outcomes in certain age groups with COVID-19 infection for a lifetime risk post-vaccination may not be a good idea.

So how could we think a system for virus detection would look? Well, it needs to be a global system, and it needs to cover both developed and developing countries. The most obvious sampling posts are hospitals and communication hubs. All people entering/passing and displaying any viral disease symptoms could be tested, and bacteria and viruses could be isolated. It really does sound like a gigantic operation, but I am quite convinced mankind would like to avoid another COVID-19. Then, the data and the samples need to be collected to a smaller number of hubs, and finally, there needs to be an International Disease Control Center, perhaps under the auspices of the United Nations. Weekly and monthly data need to be published without pollical or religious redaction, and there needs to be a committee deciding on what to advise.

Prospectively, we need to develop a series of antivirals that could be used against dangerous infections. Often, the infection per se may be looking harmless, but the immune response which follows attacks the tissue and, like we have seen in the case of COVID-19, it can quickly destroy vital organs. Therefore, remedies that can quell the immune response need to be developed. For COVID-19, we have seen antivirals, cortisone and IL-6 antibodies showing some effect but may not be fully optimised. Other principles like anti-bradykinin, inhibitors of galectin-3 or DHODH inhibitors may show greater usefulness. We also need test systems to identify individuals at risk early and taken better care of.

3.4   A Life Cared For

3.4.1   Lifetime Needs of Care

The purpose of this chapter is to provide an overview of the health care needs which are common at different ages. The thesis is that health care needs to be organised around these special age-related needs. This is nothing new, but apart from the principle of organ related and technique specialised care, it does prevail as a basic principle.

3.4.1.1 Conception, Motherhood, Pregnancy and Birth

Childbirth is risky business. Even though human like monkeys have been identified from 3 200 000 years back, and we have in most countries radically reduced both mother and child mortality and morbidity, women and children still die during labour or perinatally. Pregnancy diabetes, pre-eclampsia, bleeds and clots are identified risks, as for the fetus, asphyxia, bleeds and infections are known. For the most part, the protection against these threats can be organised away, and although zero tolerance may be proclaimed, and singular accidents may occur. Depending on the health care system and insurance coverage, there may be vast differences between different populations, races and geographies. For example, a highly educated black female in the US will have a 263% increased risk of childbirth problems than the least educated white female (REF). In developing countries, access to trained surgeons may be a huge determinant of early life health.

3.4.1.2 Childhood

Childhood morbidity and mortality are characterised by viral and bacterial infections, accidents and infrequent malignant tumours and rare inherited diseases. Vaccination against childhood diseases has been extremely successful in reducing morbidity and mortality. Access to food and clean water sets major health outcomes for large portions of the developing world. Again, socioeconomic factors and access to care determines health outcomes in western countries.

3.4.1.3 Adolescence

Adolescence is characterised by accidents, sports injuries, intoxications, suicide, infections and rarely tumours.

Some diseases like asthma and diabetes occur and may be difficult to treat in this group of patients because of denial and general problems of adaptation.

3.4.1.4   Adult Life

Adults show the whole spectrum of diseases from autoimmune, endocrine, cardiovascular, pulmonary (asthma and COPD) as well as organ based (skin, liver, pancreas, kidneys, eyes, cerebral). Work-place accidents and traffic accidents add to the mix. Tumours and infections increase over age. Socioeconomic factors, psychosocial factors, physical training and nutritional habits, abuse and choice of sports activities determine the health status

3.4.1.5   Retirement

In retired life, it becomes obvious how well you treated your parents, and how well you have taken care of your body, and whether professional life has been hard, and your financial setup generates the canvas upon which health related events play out. Generally, there is an increase in morbidity and mortality from the 60s through the 90s. Abuse and loneliness add risk and family, friends and a good economy and insurance coverage reduce risk. Diseases could be anything from osteoarthritis, to diabetes, COPD, and cancer.

3.4.1.6   Aged life

At some point in life, we will only with difficulty take care of ourselves on our own. This becomes a financial and organisational problem, and will be solved differently in different countries, depending on tradition, culture and social systems. Health care for the aged needs to be differently organised compared to health care for younger people. All disease does not need to be treated with a life-prolonging intention, and many things can be treated with symptom relief rather than hectic and dramatic interventions like surgery. Eventually, we will die from something, be it an infection, a cardiovascular event or some variant of cancer. The goal should be to keep the patient pain free and in good spirits. Overmedication and long medicine list should be avoided.

3.4.1.7 Death

Death is the great equaliser. As far as we know, no one can bring their riches in this life to the afterlife if there is one, and though the circumstances of death may differ, most health care systems do not care beyond death. One exception is the use of body parts for transplantation. This is a communication matter, a legal matter and an organisational matter to allow organs to be harvested and reused to the benefit of patients with failure of vital organs.

3.4.2 What Determines the Need For Care?

3.4.2.1 Biology vs Psychology

The extremes are pronounced where we easily see genetic disease manifesting early in life as an example of biologically driven disease. On the other hand, we can perhaps recognise that psychiatric disease emanates from dysfunctional families with child abuse, where the exogenic factors seem to predominate as another extreme.

The space between will be a mixture of nature and nurture.

Many diseases develop over more extended periods of time, and there is often a patient's delay before seeking professional help. If we are talking about proper recognizable organ-disease, there will be ample clinical signs and often laboratory derangement. This identifies the patient as belonging to a category where specialist care often is the starting point like for diabetes mellitus, and where the care will be a mixture between specialist, nurse-based care and general practice.

Psychiatric diseases are increasingly recognised as central nervous system diseases where the organic disease component over time becomes better understood.

As always, individual adaptability and degree of selfcare, one's own attitude and family support, as well as access and closeness to health care and quality of health care, determines both health care usage and therapeutic results. For serious, fast progressing disease, like some very malignant forms of cancer or in severe infections, there is little the health care system can accomplish, and care will need to focus on quality of life and supportive care.

We posit that for any given population, the size of the population, degree of urbanization, the age, gender mix, the genetic background, endemic bacterial and viral disease prevalence, industrialization, transportation system, and access to health care and health care maturity, and access to medicines will determine health care resource use as well as therapeutic results. Additional factors which determine health care resource usage are screening for disease and access to effective prophylaxis or early intervention.

We also posit that the possibilities for early disease detection and the possibilities for prophylaxis have never been greater. These possibilities are most likely to increase, as the understanding of the immune system in terms of augmenting or reducing its' activity and in simple terms decreasing the same, are on a vastly sharp rise.

Therefore, the line between nature and nurture shifts over time as we understand more and can intervene more specifically and effectively. Diseases or altered states not really classified as diseases can become possible to intervene against. Therefore, the demand for health care or health care type of interventions is likely to increase as we become more used to comfortable interventions; this is what is sometimes seen as divine and sacred life. Take a look at the plastic surgery industry and the types of interventions like Botox and hyaluronic acid injections and even tattoos that keep growing every year.

3.4.2.2 Access vs Non-Access

In any society, the question of access and non-access is key, and probably even a rudimentary system with great access is better for the majority of people than a system of great specialism and little access. For most people, a good personal economy, living in the right part of town and having a stable family and a good spouse and good friend environment is much better medicine than any known disease prophylaxis.

Since health care is highly regulated and there are high demands for safety and quality, most health care related matters are high cost and when it comes to treating real disease like cardiovascular, autoimmune, neurological, skin, eye, or cancer, the sums of money needed to cover the diagnostic and therapeutic interventions soon become unaffordable for most people. If we believe that we should strive to become less encumbered by disease during our lifetime without having to be very affluent, we need to consider how to best organise health care and how to make sure that the advancements in biotechnology and individualised approaches can reach the many.

3.5 New Therapies, But For Whom and How Many?

New therapies are being constantly innovated, and nowadays, mainly because of payor and pricing paradigms and criteria, they are focused on conditions that can yield outcomes in severe disease, so that high margins and high prices can be achieved. At the same time, we have had a period of successful expansion of the off-patent generics pharmaceuticals industry, which in many parts of the world have made medicines that were previously covered by patents and highly priced, now priced at accessible levels. Third world may be different because of corruption and monopolies, and there, paradoxically, even genericised medicines may be costly.

Given the force fields acting on the pharmaceutical industry, there may be a mismatch between the needs of the population as a whole, and those of specific subgroups with a hitherto untreatable disease, where new biological insights lead to potential remedy or even cure. And as has been explained previously, there are many health-harmful trends increasing which may better be countered by non-medical measures. So, there are several diseases that afflict a large number of individuals such as diabetes, COPD and osteoarthritis and various muscle pain symptoms and diseases affecting vision and hearing that are being less in focus for the pharmaceutical industry. The notion of market access has led to focus on more severe and life threatening and orphan diseases, where the medical need is undisputable and sometimes the disease mechanisms are better recognizable. This means that disease afflicting large number of people and where the demand for large studies increase, the likelihood for market access, pricing and reimbursement (MAPR) is reduced. All of this since allowing a new medicine onto the market at a high price and being used by many would put a lot of strain, especially in countries where health care is provided by the society. Thus, the market forces for pharmaceuticals may not always be directing resource and innovation towards diseases of the masses. Biological insight and market forces may eventually correct these imbalances.

The main problems for the development of new therapies are the slow path of novel therapies through the testing phases in man, the regulatory approval process, the time for achieving pricing and reimbursement, and then the time for adoption of the new therapy in the medical community. Additional problems are those of access based on financial and geographical factors, and insurance and payor systems.

Of course, a chronic therapy will need longer trials and if the study design says 52-week exposure, those 52 weeks cannot be sped up. The regulatory process, although lots has been done to speed up will take its time since the amount of data presented to the agencies is massive. Some areas like the speed of recruitment in clinical trials can be accelerated by using more centers and paying the investigational center more per patient.

Innovative solutions aiming at and maximizing health care quality and access across indications, across geography and across the populations over the life span of people need to consider many levels and layers in order to be effective. No single player or single payer owns the full solution in any system. And there is no single model, because during the next 20 years, we will see so many new diagnostic and therapeutic options and so much change of demography, and world economy, that we need to adapt and change.[11]

3.5.1 The Highest Health Care Cost at the Very End of Life

The more successfully we manage health during adult life, the higher the proportion of older members of the society. Eventually, we will die either from accidents and catastrophes, infections, cancer, cardiovascular disease or advanced disease in liver, kidneys, GI tract, CNS, or disease of the musculoskeletal system. Because all we know is that we have to die from something.

Most people want to avoid an early death, while we still enjoy bodily agility and while we can eat, have sex, enjoy social life, move, and travel. If disease strikes and if we are lucky enough to survive, we often find the medicine cupboard filling up, and visits to the hospital get more frequent. Often the last period in life is medicalised and sometimes hospitalised, and relative to early life this period, from a care cost and medicine cost perspective will be the most financially demanding. This is probably in no small extent due to exaggerated belief in what the medical profession can accomplish.

What mechanisms should be employed to allow access to meaningful health care while not medicalizing normal aging and end of life? This area is full of tricky philosophical questions and subjective determinations.

---

[11] Nicole, L. M. (2016). Global Population Age Structures and Sustainable Development (pp. 1-21, Rep.). New York, NY: United Natons, Department of Economic and Social Affairs.

The overall quality of care and to be able to be cared for in a decent and respectful fashion is probably more worth than attempting all diagnostic and therapeutic alternatives.

3.5.2 Are Most Symptoms and Inflictions Self-healing, and Unwellness to a Large Part Psychological in Adult Life?

As a physician and especially so in general practice, search for medical attention, diagnosis and therapy is often felt to be precipitated by anxiety and lack of personal or financial safety and security. And honestly, most conditions for which people seek medical attention are self-healing, be it respiratory tract infections, gastrointestinal disturbances, or musculoskeletal problems.

Medical education is directed towards finding patterns suggesting proper organ disease or immunological disease, but only a small fraction of all patients have a disease that needs major diagnostic or therapeutic intervention. And not so seldom do physicians miss insidious diseases like cancer, because it is a master of stealth.

Could programmatic screening, using circulating tumour markers in the blood, or circulating tumour DNA, paired with repeated imaging of for examples, breast, lungs, and endoscopic examination secure that we are not missing cancer and finding it in an early stage, so that we can look for a cure rather than in most cases ineffective treatment of metastatic cancer. Metastatic cancer is for most patients a death sentence.

3.5.3 Sport Induced Injury Can Cause Long Term Effects

We tend to believe that sports activities are on the large beneficial. And most would probably agree, but injuries in younger years do give echo also at later stages in life. Obviously, protective measures such as wearing helmets and other protective devices may help, but these are often insufficient, and injuries based on sports activities are only fully avoidable if these activities are abstained from. Should we then refrain from sports? It is probably not, but a careful analysis of which sports confer the highest number and worst injuries and long-term consequences should be performed on a society level. Particularly damaging activities should be advised against and systems should be developed to avoid these injuries that will cause suffering through life.

4 New Health Care System Paradigms

Current health care systems are basically passive receptors of patients showing clinical manifestations of disease processes, that in many cases started decades ago. This means that i) the disease has progressed from incipient to a manifested chronic disease where the normal organ/tissue architecture and organisation, as well as the organ function, has become impaired ii) there may be secondary bystander organ derangements, and iii) therapy needs to be quite drastic and forceful and may elicit major side effects and iv) the patient may need hospitalization and even intensive care and costly medicine. So, both from a suffering point of view, a therapeutic success likelihood, a cost of therapy point of view, and long-term morbidity and mortality view, the passive stance of the health care system is not optimal.

So, with that premise, how should a health care system be better designed and organised? With the new genetic and other detection tools that frontline science uses, we can now move much closer to what causes disease and which processes are specifically active in a certain individual. Cell genetic content, protein content and mRNA and other RNA content can be mapped, and forceful computer tools can map and describe the cellular machinery in quite some detail.

For patients with manifest defined disease falling in the known categories of organ disease (be it heart, lung, liver, kidneys, brain, skin, striated muscle, GI canal, ocular, auditory or endocrine system), we now have the possibility to understand the genetic background and the faulty cellular and immune system machinery to a great and increasing extent. This may allow us to intervene specifically and to reset the immune system or even to change the genetic setup of discrete organs. We may also be able to perform organ transplantation using a genetically modified organ so that there is no need for immune suppression or we may be able to skew the immune system in a tolerogenic direction with a similar beneficial result.

But of course, in the best of worlds, we should intervene before there is manifest organ damage and need for organ renewal. By knowing the genetic makeup of vast numbers of individuals, we will be able to conclude which are the factors involved in a specific disease, and over time we will learn who is vulnerable and how we may protect or intervene early before manifest organ damage.

Given a number of factors, such as demographics, poverty rates, industrialization maturity, the standard of dwellings and food intake, freedom from natural catastrophes, freedom from epidemics and status of vaccination as well as road standard, car standard and level of criminality, level of alcohol and drug abuse, we posit one could model the health care need in a given health care system setting. This will provide a starting point. And the design of the health care system can be modeled and tuned so that we get the best outcomes for the least amount of resources. If we knew the genetic setup and a few other facts like the presence or absence of certain viral infections and the immune state of an individual, we could potentially intervene earlier in the course of the disease. This could lead to pre-emptive health care.

Obviously, from a text line in a book to practical implementation to a pre-emptive health care system, there is a long and winding road, full of obstacles.

One of the main problems that come when passive health care moves to pro-active and pre-emptive, is when resources need to be derived from the passive system to an active system. The pre-emptive system necessitates investment in screening a large number of subjects long before the disease is manifest. This will seem like high cost and the benefits initially are likely the become visible long in the future. Who should bear the cost of yearly health screens, which may include early detection of phenotypic change (obesity, anorexia, muscle strength, cerebral function and humoral features) signalling unhealthy [prediabetes, CV risk factors including blood lipids, cancer screens like circulating tumour DNA etc], as well as imaging based early detection [mammography, spiral CT lung cancer screen, bone mineral density measurement etc.]), endoscopic based screening (gastroscopy and colonoscopy)? On the other hand, when we look at what we are prepared to spend on our cars to keep them going, shouldn't we be prepared to spend a similar or higher amount if we could be reassured that we will not suffer cancer, cardiovascular disease or diabetes? But it is still too early to give such guarantees, because the test systems and variables are not fail-proof – yet.

If there is a lag phase between the early detection and the time point when this class of patients would be using health care resources, then the argument for screening in the short term often is trumped by the wait and see attitude.

However, during the next ten years, one can envision that both biomarkers and imaging may become much more sensitive and precise and that biological understanding may rise to such a level that a pre-emptive active health focused system will be able to beat a passive wait and see system.

So, how can we already now start to build this new paradigm?

What would happen if we dramatically refocus the health care? To make the health care system truly based in deep biological understanding and to focus on pre-emptive and preventive medicine. Will such an investment in turning the health care system to early prevention and preemptive detection and to prophylaxis and early intervention, pay a dividend in terms of lower overall costs while attaining a higher level of health in the future? We believe it will.

4.1   Organ Repair Preparedness

4.1.1   Setting the Scene by Banking Healthy Stem Cells

When we are born, one of the first things that happen is that one of the life essential organs during gestation is being removed, the placenta. In terms of the proportion of body weight, it is by a large measure the biggest surgery most people will experience during their lifetime. The removal of the placenta and the umbilical cord represents removal of more than 1/3 of the body weight. The organs have been essential for oxygenation, nutrition, detoxification and immune protection. And now we, in most cases, throw the tissue away. Some companies and some health care systems have understood the treasures hidden in these organs, and harvest, for example, umbilical cord stem cells. These are pluripotent so that

61

they can be induced to develop into many organ typical cells.

Consider if we would routinely collect and bank these cells and keep the cells so that we can access them the day when one of our organs gives up or when we need a new bone marrow. Potentially we could be our own kidney donors. We believe all newborn babies should have their umbilical cord stem cells banked. This is a simple insurance for the future. And future technologies may be developed for new uses that we cannot even imagine now.

4.1.2 Genetic Sequencing to Reveal Danger Patterns in the Genome

Some inherited disease is not detected or evident at birth. Latency walking or talking or deficient seeing or hearing may be the first signals that something is wrong. Some of these diagnoses are picked up in the womb by specific tests, such as tests for spina bifida or mongolism. Increasingly, we may be using genetic engineering to correct these defects, and maybe even restore a healthy life.

All newborns should have their genome sequenced after birth so that genetically induced morbidity can be detected early and that active programs can be installed to intervene before organ damage is irreversible.

Obviously, many ethical and secrecy problems need to be solved, but the potential wins for these children and their families should be the driving force to make this happen.

4.1.3 Creating Mini-Organs Fitting with Risk Pattern of Organ Damage

By collecting umbilical cord cells and by sequencing the genome, we may i) learn much more about risk patterns in the genome and ii) we may be able to develop a preparedness to counteract future organ damage by a) finding early intervention stratagems and b) to use the stem cells (potentially after genetic engineering) to create mini organs which are resistant against the factors that caused the primary organ to be susceptible to damage.

4.1.4 Creating Personal Avatars

A further use of genetic information and stem cell access is to create personal avatars, which could enable very personalised therapies within silico testing of drug interventions or in vitro testing. This could promote the use of effective and minimally destructive medical therapies and be of huge value should severe disease arise.

4.2 Screening, Early Prepared Work-Up, Prophylaxis and Biosensing

4.2.1 Screening for Cancer Using Circulating Tumour DNA or Other Blood-Based Markers to Find Disease Early

New methods to detect cancer include testing for small amounts of circulating tumour DNA. Tumours shed small amounts of DNA that are different in their sequence compared to normal DNA. Our non-tumour DNA contain nucleotide (a sort of molecular LEGO which connect the two strands of DNA) base pairs coding triples that code for amino acids, defining proteins, which are recognised as self. Cancer mutates and therefore changes DNA sequences compared to normal, and therefore becomes detectable as a marker of cancer. There is a major quest for biomarkers that can be measured in blood or in excreted fluids like urine or biomarkers detected in exhaled air. We can assume that the evolution of these, not so invasive tests, will continue, and that it may possible to find tumours by screening, and to find them much earlier.

In other disease, the quest is ongoing to find early signs of disease and to find markers telling of prognosis or telling of probable response to therapy. We can believe that these tests will become more and more sensitive and more accurate in their ability to inform physicians and the health care system about the disease and to guide diagnostic attempts and therapy. The combination of thorough knowledge of the genome of the individual and the use of sensitive biomarkers will allow a deeper understanding of disease and disease drivers, and help therapy to become more suited to the individual.

Obviously, this added clarity and granularity together with AI based information and decision support systems will be of tremendous help to physicians and patients.

However, and this becomes extremely important, this means that the current passive approach of the health care system to be the last resort of help at times of desperate need and emergency may become something of the past. Or at least less predominant. Catastrophes will still occur, but with the new individualised knowledge we can turn the health care system 180 degrees so that we move towards pre-emptive health care, that is fast on the ball. And where we spend less resource on fending off unnecessary futile anxiety driven health care and focus on the individuals who need appropriate diagnostics, medical or surgical therapy, and for some subjects what we can call a bio-system correction. That is DNA based correction or cell therapy-based correction or immune system reset via other mechanisms.

4.3   Diagnostics

4.3.1   AI Enabled Systems as First Port of Call to Elucidate Symptoms and Real Reason for Seeking Health Care

Most visits to the health care system results in very little detection of what we call real disease and is often precipitated by anxiety. One of the main tasks of any health care system is to categorise patients correctly so that serious and real disease is not missed and that symptomatic states without the underlying disease can be quickly reassured with minimal cost and use of a resource. It may be possible to use artificial intelligence as a mechanism to classify patients correctly.

4.3.2   Full Fast Diagnostics

We don't want delays in detecting disease and finding a cure. Modern methods of non-invasive diagnosis (read imaging) and semi-invasive methods (read blood based or sampling from other fluids), or invasive (needle biopsy, or open surgical biopsy) can quickly establish the cause of many diseases. However, there are many examples when a diagnosis is missed or when patients are misdiagnosed. In any case, we want a fast diagnosis and would rather submit to more tests if that could increase the likelihood of achieving a diagnosis. The utopian concept of 1-day diagnosis and 1-day cure is what we must strive for.

If the health care system can quickly sieve of the not likely sick, then more resources can be spent on the individuals with real disease. For these patients, it is vitally important to go broad and quick and to allow multiple diagnostics approaches in a short period of time. Thus, medical diagnostics centers can be envisioned where the full machinery is at hand to go from DNA, to investigate the metabolome, the proteome, the endocrine systems, the immune system, the cardiovascular system, the pulmonary system, the blood forming system, the CNS and the peripheral nervous system, the microbiome and the virome. And where sampling from tissue and staining with multiple probes can happen quickly. This focused concentration of technologies and with the use of bioinformatics and artificial intelligence can turn fast insight into fast therapy and faster recovery.

4.3.2.1 Full Individual Pre-Characterisation, Directed Prophylaxis and Fast High Sensitivity Intervention

When we move through life, our likelihood of disease changes (from a small child to adolescent or adult there is a decrease), and in later life likelihood of disease increases quite evidently. If we consider a system where we are already well characterised genomically, we can envision a system and a situation where the primary understanding of the possible basis for disease is better known and characterised and where we already have employed directed prophylaxis to certain individuals who have genetic setup known to predispose to certain disease. These individuals should be followed more intensely and there should be readiness for intervention.

4.4 Therapeutics

4.4.1 Going Fully Personalised by Bringing Biological Research to Bedside

We can now sequence the full genome and the protein expressing part, the mRNA portion in a day. Our understanding of human immunome (the immune system) will develop so that we can detect the antigens to which humoral and cellular immunity are set to react, and describe self-enforcing loops and tolerogenic loops, both of which may be harmful in the specific disease setting. We may also use immune avatars and in-silico modelling in order to understand how to best intervene. With these tools we will be able to take the step into fully personalised therapies, where we can create patient specific therapeutics which may effective stop the immune system to overreact to certain antigens and tissues, or in other cases, entice the immune system to find and kill pathogens and cancer which try to evade detection and destruction.

4.4.2 Modelling Individual Response Before Individualised Therapy

4.4.2.1 In-Silico

With more meaningful, detailed knowledge about disease and immune system, we will be able to create disease models and do more testing of disease versus therapy will be done in silico. With a systems biology canvas, we can create individual disease models which then can be tested to create actionable insight into the choice of therapeutics and combinations of therapeutics. We can even think of using medicines that have been approved for one indication but where we know the biological action in full, and thus may be able to use in other indications and in novel combinations. In silico testing may be able to model these interactions and may lead to individualised therapy.

4.4.2.2 Organoids

Another way to individualise therapy is to test medicines on person derive organoids. With three-dimensional cell culture (3D cell culture), the cells behave more naturally than when cells are cultured in a flat monolayer. The cells begin to show features that more closely resemble the actual cells in a live organ (e.g., beating heart or lung tissue with ciliae). These organoids are additional means to create individualised therapies and to test what may work in the whole live patient.

4.4.2.3 Super Individualised Care

Combining super detailed mapping of genes, cells, and cell machinery will enable us to pinpoint defective parts and create settings where we can intervene without creating toxicity in the patient. This is in stark contrast to what is done today, when doctors try new therapies with very limited knowledge as to how the patients will react, both from an efficacy point of view and from a side effect perspective. And the one would ask whether this will be overly expensive, and surely, it will be expensive to begin with, but as with all technology, costs will reduce, and success rates will go up.

4.5 New Ways to Organise Health Care

Once we start thinking of how to optimise health care it is not difficult to think about other aspects leading to major shifts in the health status of the population. This chapter looks further into the organisation of health care.

4.5.1 Organisation and Priorities

4.5.1.1 Create the Best Possible Starting Material Through Massive Investment in Childhood Physical and Psychological Health

Let us consider society as a producer of 1 product: healthy 20-year-olds. Arguably, a society which produces very healthy 20-year-old individuals, is more likely to keep the cost for health care lower than a society that fails in this respect.

How is this accomplished without resorting to methods associated with the 3rd Reich? This ambition leads to active intervention in utero where we already accept to abort fetuses with Down's syndrome, and one might consider screening for odd genetic disease and again use early abortion as a method to decrease the number of subjects affected by these disorders. It may seem cynical to some to state that subjects with intractable genetic disease will absorb large proportions of health care resources. To some, this kind of reasoning sounds cruel. Maybe, but of course to evolve and devolve functionally in your early years also sounds very cruel, as is condemning a family to a life bound and limited by a severely handicapped child. The life trauma of seeing your beautiful and beloved child dying cannot be overstated.

So, society must set a very clear target to achieve very healthy 20 year-olds, and both systems to enhance the health and to avoid handicapping young people must be installed.

With the obesity epidemic, much more focus must be put on getting children moving. Sports activities with a low risk of permanent injury must be promoted, and protective measures must be installed in the more accident-prone sports.

4.5.1.2 Use Screening and Prophylaxis for Adult Productive Population

Some parameters become very important for a healthy population, such as body weight. Following body weight and intervening is one of the most fundamental and important health care interventions that can be made.

69

In school, focus should be given to corporal and mental health and body exercise and movement needs to take a huge part in order to counteract the overall tendency for identification.

From a certain age, and this will depend on a number of factors such as financial, all adults should undergo a health screening to detect incipient organ disease and cancer. The goal should be much earlier detection of severe but treatable disease, and cancer before becoming metastatic.

4.5.1.3 Industrialise and Professionalise Care for Real Chronic Disease

The health care system as is and with its very widely defined role of both caring for our health nuisances and for major disease and for health catastrophes, has an impossible task, similar to combining McDonald's, with a luxury restaurant such a the Danish Noma and with a fire brigade, always ready to act in case of emergency.

We believe that industrializing and further professionalizing the care for patients with clearly defined diseases such as autoimmune disease or severe organ diseases such as heart or liver disease, and securing the best possible.

4.5.1.4 Establish High Throughput Centers for Immediate Cancer Care, and for Disease Where Time is Critical

There are few more threatening disease-diagnoses than cancer. Patients with cancer should be treated immediately and get immediate full workup so that lag times and delays should be avoided.

Other diseases which deteriorate quickly include multiple sclerosis and rheumatoid arthritis. These should also be given priority as functionality may suffer if adequate therapy is delayed.

4.5.1.5 Focus Care of Elderly to Prophylactic Muscle and Balance Training and Nutrition

Health care for elderly can be made very complex. Often elderly end up with cupboards full of medicine, and no one knows why these were given in the first place, and no one knows which ones are meaningful. Overmedicalisation should be avoided.

The real determinants for:

4.5.1.6 Creating a Novel Health Care Diagnostic Experience via AI, Immediate Multiomics, Imaging and Diagnostics Evaluation Teams

With the fast pace of development of molecular dissection of disease and genetics, no single physician can keep up to speed. Also, the doctors of today and yesterday are not equipped to compute the data which can now be gathered and to fully make use of such a detailed and complex narrative. We suggest setting up specialised teams of diagnostic and therapeutic experts in order to fully depict and fully model disease and therapeutic interventions. These teams will need both experts in biology and in genetics and in disease modelling, AI and in therapeutic modelling. Initially, the throughput of such teams will be low, but over time more and more will be learned and with focused teams that chase to detailed diagnosis and to the most meaningful intervention will shorten.

4.5.2 Novel Professional Categories

With a novel setup of health care there will be a need for new professional categories, which will include experts in certain tissues, and in certain types of biology, including immunology, and in bioinformatics, and modelling. These professionals will learn to work together with the patient in focus in order to most effectively elucidate biology, diagnosis and therapy.

4.5.2.1 Smell Based Detection Systems

It has been known for long that certain disease states create recognizable smells. We are not as sensitive as dogs and only detect a small range, but clinically the smell of acetone from diabetes mellitus with ketotic hyperglycemia is taught in medical school and recently, it was written that patients with Parkinson's disease smell in a characteristically acrid way. Another example is foeter hepaticus from patients with liver cirrhosis.

New technology uses nanosensors and this can add to the detection range and specificity of diagnosis of CNS disease (Parkinson's disease, Alzheimer's disease, multiple sclerosis), cancer (e.g., lung cancer) or lung disease (asthma and COPD). In asthma, there is already use of devices that measure exhaled nitric oxide, which is increased when the asthma is not well treated.

4.5.2.2 Heart Monitoring Real Time All the Time

The ticker is a very good friend as long as it works flawlessly, but anything even slightly serious with the heart soon becomes very life limiting. Many times, heart problems are detected as heart rhythm disturbances but other signs include breathing difficulties or chest pain. Once defined as a heart patient heart rate, heart rhythm, ECG read, oxygenation becomes very central to the monitoring and finding early signs of disturbances may lead to early intervention, which may be life-saving.

With modern technology, we can monitor several of these parameters in real time, and this may allow early intervention. The Apple watch can transmit an ECG proxy. Another product is a chest borne technology called SPYDER, which transmit to the cell phone and which relays to a reading center with 24/7/365 surveillance. This is just the beginning and other real time monitoring/intervention can be added. Other measurements such as oxygenation, breathing rate and pattern, the activity of the stress nervous system and blood glucose are other parameters which when monitored every day for a large number of people will make pattern recognition possible so that automated detection can warn and intervene to avoid bad health outcomes.

Pacemakers have developed to life-saving intervention machines. They can now counteract a number of heart rhythm disturbances, including ventricular fibrillation (when the chambers of the heart just contract entirely irregularly and create no effective pump action) is life threatening.

4.5.2.3 *Blood Sugar Monitoring and Automated Systems*

New technology uses continuous read of blood glucose and algorithm based/automated injection of insulin (artificial pancreas) gives better control of diabetes and may offer a reduction in episodes of dangerously low blood sugar (hypoglycemia) or out of control high blood sugar (hyperglycemia).

4.5.2.4 CNS Monitoring and Electro-Stimulation

CNS disease can be influenced by chemical medication or by sending electrical impulses to certain parts of the brain. Patients with tremour (read Parkinson's disease) or with epileptic attacks may benefit from electrostimulation.

5   So, What About the Future, Then?

Our life expectancy is increasing, and our expectations of life as well. Still, 4% of all babies are born with severe malformations and the risk of dying as an adolescent is about 12 in the best European countries and above 28/100 000 in the USA. In less developed countries the figures go up dramatically. Mortality between 15 and 60 years of age lies at about 70 in the best EU countries, and at 114/1000 inhabitants in the USA. In countries like Denmark, the adult mortality was about 100/100 000 at the year 2000 and in 2018, it was measured to be 65/100 000 inhabitants between 15 and 60 years of age. USA had 114/1000 in the year 2000 and went down a bit and then up to 114/1000 again in 2018.[12]

So, life will differ depending on where and when you are born, and generally speaking, we will get older as a species. The arbiter of life expectancy or probability of adult disease-free years is multifold and medical advances may help to some extent. The fact that we have decent dwellings and that we can count on food supply and have reduced the dangerous occupations probably control most of adult morbidity decrease, and health care and medicines to a lesser degree. The probability of living a decent, healthy life after 70 becomes more health care dependent and this trend increases over growing older.

The bigger questions are related to achieving an excellent health starting point in adult life. This may to a great extent, depend on matters of food intake, body weight,

---

[12] World Health Organization. (2021). GHO | By indicator | adult mortality rate (probability of dying between 15 and 60 years per 1000 Population) (mortality and global Health estimates). Retrieved from https://apps.who.int/gho/data/node.imr.WHOSIS_000004?lang=en

mobility and cardiovascular health, rather than the health care system. Later in life, it becomes a matter of how we can best organise and distribute health care with the best possible efficacy and effectiveness. Current systems with chaotic health care and in-effective preventive care are not sustainable.

The high tech and high diagnostic sophistication-based health care need to be employed as low-cost mass preventative care, and with great precision and quality for people with severe disease. At present, there are few signs of concerted and conscious movement towards a smart system of health care and disease prevention, rather than after-the-fact heroic interventions or high cost moderately effective medicine.

All change depends on consciousness, decisions and implementation. If we can at least start to increase consciousness and debate, maybe the rest will follow.